Genetically Modified Crops

Benefits At A Glance

SANDHYA ANAND

DEDICATION

To My Late Parents, Who Has Been A Constant Inspiration And Guidance In My Life, And My Loving Family....

CONTENTS

ACKNOWLEDGMENTS

*I would like to express my special thanks to the support from my family. I also express my sincere thanks to Dr. S. Shiburaj (TBG&RI) for his constant encouragement and guidance. Thanks to **AGRIHORTICO** for their valuable suggestions on the topic and for giving me the chance to lend these words for a greater cause.*

1 INTRODUCTION

Genetically modified crops are the crops whose genetic makeup is changed. The technology enables you to transfer the genome of one organism and allow it to be inserted to another organism. The genetic makeup of an organism specifies its characteristic features. The genetic material of all plants and animals are made of DNA which stands for Deoxyribonucleic Acid.

DNA of the organism carries the set of instruction to produce enzymes which carry out the chemical reactions which take place in an organism as a whole. These chemical reactions are numerous and are often interlinked like a complex web. Enzymes are proteins which are important in the biochemical functioning of an organism. DNA also has instructions to produce other proteins which make up the structural features of an organism.

For example, the DNA of a plant determines its height, size and shape of leaves, type of flowers, colour of the flower etc. These are visible characters which are cumulatively called as the phenotype. In short, the genotype i.e.; the genetic material determines the phenotype. Genetic modification involves modifying the genotype to change the phenotype.

Genetic modification is usually employed in crop plants to increase the economic yield either by directly increasing the production, or by increasing the plants resistance to herbicides, pests, antibiotics etc.

Most of the crop plants are productive under ideal conditions of cultivation. But in nature, the perfect conditions of growth seldom occur, thanks to industrial pollution, global warming, unexpected natural calamities etc. The effect of these biotic and abiotic stress factors is visible on the productivity of the crop. Loss of quality as well as decrease in quantity is the most common among them.

Nutrient density in soil is one major factor affecting the crop productivity. Fast utilization of soil resources, especially after industrialization has led to depletion of soil nutrients. This has in turn resulted in overuse of fertilizers to account for the loss.

External addition of chemical fertilizers and pesticides over the years has resulted in serious clonal loss of crop yields. The techniques of biotechnology have given many solutions to overcome the effects of such stress factors.

Genetic modification is done in the crop plants using biotechnological tools to produce new genetically modified GM crops. GM technology can also produce crop plants with additional nutrients than traditionally grown crops. Such plants are said to be fortified with nutrients. Crops are usually fortified with necessary vitamins, minerals, antibiotics etc.

The changes induced by Genetic Engineering technology are heritable, which means are passed on to the subsequent generations. The inserted DNA becomes part of the genome of the crop and is hence passed on to the offspring. Seeds and seedlings carry the modified genome so that the traits are expressed in subsequent generations.

Genetic modification of crop is not a new concept. For centuries, food crops have been changed through selective breeding. While genes can be transferred during selective breeding, the scope for exchanging genetic material is much wider while using genetic engineering.

Genetic modification of crop plants thus involves improving the yield quantity and quality either by the introduction of a genome for desired trait into the plant genome or by modifying the inherent plant genes specifically selected for the expression of a trait. Biotic stress factors included bacteria, virus, fungi, insects, pests, nematodes etc. Abiotic factors like soil salinity, extreme temperatures, drought, decrease in soil nutrients etc. can also deteriorate the yield quality. Genetic Modification has been found to be effective in combating with such factors which affect and decrease the yield quality and quantity.

Advantages and Disadvantages

1. *Fast Results*

    Perhaps, the greatest advantage of genetic modification of crop plants is the ability to observe the results of experiment sooner than what a conventional breeding process would normally take. Traditionally, farmers have resorted to breeding practices to improve the yield ever since agriculture. The chances of failure were also much higher.

2. *Exact Desired Traits*

    Genetic engineering techniques can identify isolated and transfer the gene for the desired trait into any organism of choice. The desired gene is selected either by constructing gene libraries from total genome or by cloning the DNA fragment by PCR technique. Gene for transfer might be taken from cDNA libraries or total genome libraries. Genomic DNA libraries on the other hand consist of all the DNA sequences from its organism.

    Crop plants are genetically modified for developing traits of economic importance such as,

    ✓ *Herbicide Resistance*
    ✓ *Viral Resistance*
    ✓ *Antibiotic Resistance*
    ✓ *Pesticide Resistance*
    ✓ *Resistance To Insects*
    ✓ *Drought Tolerance*
    ✓ *Tolerance To Extreme Temperatures*
    ✓ *Fortification With Nutrients Such As Vitamins And Minerals*

Marketing of Genetically modified foods have been without proper markings in the earlier times when they were introduced. Though they occupied a very small percentage of the total marketed foods when they were introduced in US in 1997, now they form approximately 67 % of the total processed foods marketed. Rotting resistant tomatoes (Flavr Savr) which were introduced later were soon followed by a variety of processed foods, all without proper labelling. The scenario soon changed to include GM crops of economic significance such as Bt Cotton.

Currently, the five major countries which grow GM crops *(in their decreasing order of area of cultivation)* include

1. *United States*
2. *Brazil*
3. *Argentina*
4. *India*
5. *Canada*

The unprecedented growth of GM crop cultivation is rather alarming. As on 2014, 181.5 million hectares of the land is used for cultivation of GM crops worldwide. Of this, 90 % of area falls within the borders of the five major countries. In 1994 when the crops were just gaining momentum in the market, the area was just around 1.7 million hectares.

Of the GM crops grown worldwide, most prominent ones are soybean (83%) and Cotton (75%). Though in a minor proportion, Maize (29%) and Canola (24%) are also being cultivated.

Though India was late in allowing GM crops to be grown, it soon gained the first place in production. The Bt Cotton variety developed by Monsanto was first approved for commercial cultivation in 2002. By 2013, the cotton production almost tripled. The final yield also saw a considerable increase up to nearly 100 %. In fact, almost 95% of the produced cotton in India owes its dues to the GM seeds and made India the leader in cotton production.

New-generation GM crops are now also being developed for the production of recombinant medicines and industrial products, such as mono clonal antibodies, vaccines, plastics, biofuels etc.

2 HERBICIDE RESISTANCE IN GM PLANTS

Herbicides are used by gardeners and farmers for weed control. Techniques of genetic engineering allowed researchers to isolate resistance genes from other organisms, introduce and incorporate them to potential susceptible crops. Although these Herbicide Resistant Crops were originally developed using traditional methods, they are now being produced using genetic engineering methods. In fact, genetically modified Herbicide Resistant Crops account for 85 % of the total are of cultivation of GM crops.

Benefits of Herbicide Resistant Crops include improved farm management practices and an increase in total yield. Glyphosphate resistant crops such as soybeans, cotton, canola and corn made significant changes in weed management practices adopted in traditional farming methods. Farmers resorted to mono herbicide application since the introduction of genetically modified Herbicide Resistant Crops and the subsequent yields were quite significant as established by studies for more than a decade since their introduction.

However, this practice turned unsustainable on the long run since weds too developed resistance. Multiple Herbicide application was needed which made farming less cost effective and hence an unattractive profession. In order to overcome this challenge, scientists are now searching for effective solutions in development of multiple herbicide resistant crops and have succeeded in a few cases.

Possible long term implications on human health and environment, the risk of creation of herbicide resistant super weeds, monopolising seed stocks of crop plants by agribusiness lobbies and threats to the biodiversity conservation efforts through sustainable farming methods are just a few of the major issues raised by the opponents of Genetic engineering technology. Last, but not least, the questions concerning ethical research practices have also been on the rise.

Physical removal of weeds can be cost demanding and often increases the capital cost for farmers. Traditional methods such as tilling can prove to be a time consuming affair too. The original reason for spraying herbicides has been to mitigate such effects. But the process requires a lot of calculation and patience since the amount of herbicide needed can always be prone to judgemental errors. Due care has to be exercised so that the herbicide does not affect the plants selectively killing the weeds. Genetically engineered crop plants addressed this issue by their selective tolerance to herbicides. This ensured that farmers calculated and applied the right amounts of herbicide depending on the weed quantity. Use of lesser quantities of herbicides also helped to protect the environment from the potential damages due to widespread use of herbicides.

Monsanto's Round up resistant soybean is one such herbicide resistant crop which requires just one round of application of weed killer 'Round up'. This has helped in considerable reduction of farming costs, simultaneously reducing the use of herbicides. It also indirectly gave protection to the proportion of crop plants which would otherwise have been an agricultural waste off due to application of herbicide.

Consider the case of glyphosphate resistant tobacco plants. Mutant EPSP (5-enolpyruvylshikimate-3-phosphate (EPSP) synthase) synthase gene originally isolated form glyphosphate tolerant wild petunia plants with high level of expression reportedly increased the yield when engineered into tobacco plant. These engineered tobacco plants were resistant to glyphosphate thus allowing effective weed control without destroying the crop. Such selective weed killer resistant strains have been developed in crop plants such as corn, cotton, and alfalfa too.

Similar researches were conducted for finding selective tolerant cell lines to herbicide like Imidazoline, Sulfonyl urea, Bromoxyline, Triazine and Phosphoeinothiazine.

But incomplete resistant to glyphosphate emerged to be an important concern especially under stress conditions. This effect when coupled with less frequent applications of herbicide in farms led to reduced yield in the first F1 generation of genetically modified plants. Though the effects were limited in second generation, further studies and detailed analysis especially on the expression of herbicide resistant genes in crop plants, both singly as well as in combinations; and under stress conditions need to be undertaken so as to eliminate the concerns on GM crops.

Origin and Growth of the Super Weeds

Super weeds are resistant to the commercially available weed killers and one of the major after effect of the cultivation of Genetically Modified Herbicide Resistant Crops. Their emergence have raised the concerns of wide spread cultivation of GM crops.

The rise of super weeds was never due to GM crops solely. The emergence of such resistance in weeds has been a natural process and slower before the advent of GM crops. GM crops effectively reduced the use of multiple herbicides, thus prompting the farmers to rely on a single herbicide. (For e.g.: Round up is an herbicide marketed by Monsanto). Continuous use of single pesticide coupled with monoculture farming method has eventually accelerated this natural process eventually resulting in the rise of super weeds. Planting the same crop year after year in the same habitat thus had detrimental effects. For the lands cultivated using Roundup ready crops and Round up application, the initial cost reduction was temporary. Subsequently, it resulted in the growth of Round up resistant crops which required more herbicides than initially applied before the GM crops thus ultimately reversing the effects of Genetic modification. Some of these round up resistant super weeds include palmar amaranth, Artemisia species (commonly called as 'ragweeds'), water hemp and mares tail.

A lesser noted but important reason for the rise of super weeds is the sheer negligence of the farming community to other weed control measures.

  a. *Crop Rotation*- Altering the crops planted in a specific habitat can alter the use of herbicides required and thus effectively control the dependency on a mono herbicide. The nature of weeds for different crops also varies thus effectively reducing the likelihood of herbicide resistance among weeds.

  b. *Cover Crops*- They are effective in preserving the biodiversity of a habitat, managing soil erosion, and improving soil fertility.

7

Cover crops can also reduce weeds and preserve genetic diversity.

c. *Tilling For Small Proportion Of Weeds*- Traditional methods like tilling can be used in case the weeds are very less in quantity. This can help to cut down the cost and use of herbicides.

d. *Allelopathy*- It refers to the production of allelochemicals (Phenolic compounds, long chain fatty acids, terpenes, simple acids etc.) in response to stimuli to suppress the seed germination, growth of shoots and roots in other plants nearby. These effects are collectively called as allelopathic responses and can be used in effective weed control. Allelopathy is exhibited by both crops and weeds. Some examples include barley, rye, buckwheat, sunflower, oats, alfalfa, wheat etc.

However, the strength of allelopathic effects varies widely and is often unpredictable. For example, one crop which is strongly allelopathic to a specific weed might not be able to control the infestation of other weeds.

Super weeds caused due to pollen flow from the herbicide tolerant crops to weed plants have caused substantial damaging effects like spread of therapeutically important genes to wild plants, loss of genetic diversity, transfer of transgenes to non-crop plants, and international disputes in gene patents.

However, Governmental policies across the world and convenience have caused the farmers to neglect the natural mechanisms and resort to the use of GM crops and single herbicides. The reasons can vary from the initial cost reduction, ease of application etc. to the monopoly of the seeds and herbicides by multinational companies.

3 RESISTANCE TO DISEASE CAUSING VIRUSES, PESTS ETC

Plant diseases are primarily caused by viruses, fungi and bacteria. Genetic modification of these plants by Plant biologists aim to alleviate the crop loss and improve resistance to these diseases.

I. Virus Resistance In Plants

Virus infections in plants can be devastating. Cross protection methods were adapted in the past to improve virus resistance in plants. The method is similar to vaccination in humans. Mild strains of viruses are allowed to infect the plant which gains resistance to a full-fledged viral attack.

However, there are certain risks associated with the process. The strain might undergo mutation to a virulent one, or can be an unsuspected pathogen or another crop. It is also highly unpredictable what the synergistic association of the strain with other common pathogens can result in. Also, the gene expression pattern, virility of the strain and the severity of the infection cannot be predicted under stress conditions in the farm.

Method 1: CP Mediated Virus Resistance
Genetic engineering methods have been employed to induce virus resistance in crop plants. For example, CP (Coat Protein) mediated resistance against Tobacco Mosaic Virus (TMV), Potex virus; cucumovirus and alfalfa mosaic virus were induced using Agrobacterium tumefaciens. Pathogen-derived resistance was found to be effective in developing virus resistant tobacco by Roger Beachy and his team. For this, a segment or a

complete gene from a viral pathogen is inserted into the crop plants, which in turn develop resistance.

Apart from the CP gene, replicase gene sequences, satellite RNA as well as defective viral components that can affect the viral expression are also used. Field tests carried out in Genetically Modified Muskmelon, Papaya, Tomato, Potato, Cucumber etc. have shown promising results of increased viral resistance although the exact molecular mechanism of this CP mediated resistance is still under evaluation.

Significant increases in yield were also obtained using transgene PRSVCP and a combined ZYMU+WMV2 CP in squash plant.

CP Mediated Resistance also offers protection against vectors like aphids. Again, the detailed mechanism of action is under evaluation. In 1950s, Hawaii's Papaya yield decreased drastically due to PRSV infection. The CPMR transgenic Papaya varieties introduced in 1991 was virus resistant for a longer duration. When crossed with other varieties, it resulted in a variant Rainbow variety which had a yield almost 20 times that of the regular ones. Rep gene mediated resistance is narrower but strong enough to combat a higher dosage of the virus.

Method 2: Movement Protein Mediated Virus Resistance
An alternate approach relies on the use of Movement Proteins which are necessary for movement of cell viruses from one plant cell to others in the process of infection. The method however was effective only in cases where a single gene was responsible for controlling the movement protein expression. When two or more genes are involved; interactions between viral encoded and transgene encoded Movement Proteins also seemed to play an important role, the details of which are yet to be explored in detail.

Satellite RNA sequences when introduced in Tomato plants showed remarkable resistance to CNV infection in field trials. However, precautionary measures need to be employed since the satellite RNA spread is in an epidemic fashion.

Defective viral components/ribozymes (self-cleaving RNA enzyme) are also used to create resistance against viruses in transgenic plants.

Method 3: Disease Resistant Genes, Enzyme Inhibitors, Specialized Proteins Etc.
A secondary approach makes use of disease resistant genes, protease inhibitors, Interferon like systems, ribosomal inactivating proteins, antiviral

plant bodies etc. Systemic acquired resistance develops first at the site of infection and then progresses systematically to other plant regions.

A major obstacle in the transgenic technology is the extreme diversity among viruses. A lot of variant viral strains can result in the same disease symptoms. Hence a single gene approach might not be fully effective to combat all viruses. The combined and multiple effects of co expression of multiple genes need to be targeted for further evaluation.

In order for an effective implementation the basic functionality of pathogenesis, the variants of viruses along with their genomes and the transformation processes of all major crops along with their resistance and yields need to be explored in detail. Promising results to mitigate effects of crop loss can be possible only with such a simultaneous top down and bottom up approach.

II. *Resistance To Pests*

Insects being the most diverse and abundant forms on earth are the greatest cause of concern for crop loss too. Although, only 0.5 % among them is qualified to be considered as pests, they are substantially able to inflict a crop loss of about 20 % of the total annual yield in the world. If you consider seriously, only a minor fraction of these pests cause serious damage to crops.

Ever since mankind began to grow crops for consumption in a systematic way, the challenge of pest damage has been a major threat to availability of food. Although crops could be grown in large and confined areas, the insects too developed resistance on the natural course. Since higher yield was the primary selection criteria, the crop plants developed during the agricultural revolution focused only on the yield of crops. This has resulted in selection of crop plants with less insect resistance also to be selected if they possessed high yield traits.

Crop plants susceptible to insect infestation soon appeared leading to a competition between humans and insects for food. Suppressing the insect population was perhaps the only way to avoid severe food shortage. The tropics and subtropical regions are the primary regions where majority of the crops are grown and unfortunately the climate conditions are conducive for the growth of pest insects too.

The issue can be further devastating in developing countries. The growth of huge population of insects forces the farmers to rely on pesticides which can potentially harm the humans too. Apart from this, traces of pesticides can contaminate the water bodies and agricultural lands leading to potential loss of soil quality and groundwater too.

Pests also find new homes often, thanks to global trade and transport. Such pests when introduced to foreign soil often have less or no natural control agents to check their growth and soon outgrow the native species and infest crops more vigorously. Pesticides and insecticides have been able to control the infestation only to a certain extent due to the development of such invasive species. These insecticides have also been harmful to the pollinator insects which are beneficial in natural plant propagation. The development of insect resistant plants by genetic modification has alleviated the effects to a certain extent.

The majority of the insect resistant crop plants like corn, cotton, brinjal etc. have been genetically modified using the genes from a naturally insect resistant bacteria, Bacillus thuringiensis. The bacterial toxin known as Crythat is harmful to insect species. The gene responsible for the production of the Bt toxin is isolated from the bacteria, cloned to form multiple copies and inserted into the young plant to develop insect resistance.

GM insect resistant plants can withstand the infestations from root worm, boll worm, European corn borer, corn ear worm and tobacco bud worms. Bacillus thuringiensis was first isolated by Ishiwata Shigetane in 1901. But, it was Ernst Berliner who first identified the preferential toxic effect of the bacterium towards certain insects, especially their larval stages. However, the Bacteria were not even considered as a potential pest control method until 1928.

It took another ten years to introduce the first Bt based biopesticide known as Sporine on a commercial scale. It was made available to the farmers in France first following which wide research was carried on to isolate the new Bt strains with resistance to a wide range of insects and develop biopesticides.

The introduction of Bt resistant genes to develop transgenic plants in 1995 revolutionized the concept of pest control. The Bt bacterial toxicity is mainly attributed to the production of Cry protein which are highly target specific and insecticidal against pest larval form. The Cry proteins break down to form Cry toxins which bind to the surface receptors of cells lining the intestinal gut of insect larvae. This causes the intestinal cells lining the gut to destroy themselves and thus creating a channel for entry into the insect body. Upon reaching the insect body cavity, they proliferate and further invade the other cells eventually destroying the insect.

The Cry protein is highly specific to certain groups of insects and therefore more effective in controlling pests alone causing no harm to other organisms. Toxicity studies carried out in rodents showed no significant harmful effects. The protein was also found to get destroyed in the acidic stomach environment.

Questions And Issues Of Concern

Genetically modified inset resistant crops certainly helps to reduce the amount of pesticides the farmers otherwise would have needed. The extent of reduction in the use of pesticides and the economic gain from such reduction need to be further evaluated and properly analyzed against the cost of development and cultivation of transgenic plants.

Apart from this, whether Cry proteins undergo modification to infest new pests or cause harm to other organisms need to be studied extensively. Comparative studies between the effect of application of Bt based pesticides and Genetically modified insect resistant plants have not been undertaken much. Lastly, the safety concerns surrounding the GM crops need to be addressed beyond doubts.

GM Insects

Off late, researchers have focused on insecticide free mechanism such as creation of genetically modified insects which are unable to infest the crops. They are developed by insertion of new genetic material into the insect native DNA and allowing them to grow. In 2006, A GM pink bollworm moth strain has been released into the environment in U.S as a pest control measure. The strategies include population suppression to suppress the insect population by production of non-viable offspring and population replacement method to replace the wild strain in environment with the GM strains. However, their effectiveness and ecological impacts are still unknown.

4 ANTIBIOTIC RESISTANCE IN GM PLANTS

Antibiotic resistant crops are another important contribution of Genetic engineering. Antibiotics are the most important among the pharmaceutical products which are produced by microorganisms. These classes of chemicals can inhibit and even kill other microorganisms. Sometimes, they turn out to be harming the useful microbes. In such cases, the use of antibiotics can kill the useful microorganisms and hamper the proper development of crop plants.

To overcome this, scientists have tried and created antibiotic resistant crops by the introduction of antibiotic resistance genes from another organism, usually a microbe. These genes are portions of DNA fragments which can code for proteins to selectively inhibit certain antibiotics effectively protecting the cells from the action of such antibiotics.

Most antibiotics are produced by Bacteria. They are found practically everywhere in the environment. In fact, we are constantly exposed to the bacteria from surroundings. You must have come across a large number of offspring being produced by certain animals, fishes and insects. Out of them, only a few survive to become adults. Charles Darwin termed this phenomenon as 'survival of the fittest'. Likewise in microbes, 'survival of the fittest' strategies are employed by certain bacteria to outgrow others. Production of antibiotics is one such strategy. Often the microbes which grow in the same habitat fight with each other for food, space, etc. Usually microbes which produce antibiotics also carry resistance genes for that specific antibiotic. This is mainly to equip themselves against self-destruction. These antibiotics inhibit the chemical/metabolic processes of another organism (mainly bacteria) eventually destroying the organism itself by affecting its growth and replication.

However in the process of evolution, the targeted bacteria try to acquire the resistance genes in order to survive. Gene transfer mechanisms are simpler, random and more common among bacteria. Such acquired resistance also spread fast due to such horizontal transfer of genome among bacteria. Indiscriminate use of antibiotics has led to development of resistance also. For example, antibiotics are now being widely added as supplements in animal feeds, insecticides and fertilizers. Often they are also sprayed on crop plants as a protection mechanism against disease causing microbes.

Antibiotic resistant cells are identified based on the presence of specific markers (gene sequence) and the genome is extracted from them. The specific gene sequence is then introduced into the target cells and the resultant cells which are resistant to antibiotics are chosen for identifying transgenic ones. These transgenic cell lines are used for incorporating desired traits and form genetically modified plants.

The antibiotic resistance genes thus serve only as markers to identify transgenic plants. Even though herbicide tolerant genes are also used similarly, they serve the purpose of protecting the plants from a specific herbicide which when applied can kill the adjacent weeds alone. However, the antibiotic resistant GM plants pose a serious risk of introducing antibiotic resistant microbes also in the natural environment.

These antibiotic resistant marker genes can be either be DNA sequences intended to serve as markers for selection of genome for amplification. Or they can be just identification tools to select the cells which have taken up the desired trait. This is particularly useful since only a minor fraction of the targeted cells take up the desired genome in laboratory conditions. Hence, it is very important to isolate the transformed cells from which GM plants can be grown.

Antibiotic Resistance Markers In Use

The most common among them is the nptII gene which offers resistance to kanamycin and neomycin. The gene is selected based on its non-utility in the therapeutic field. Most of the bacteria found in nature (20-40%) are found to be resistant to the antibiotic. The resistant bacteria form a major portion in the pig gut.

These are widely favored due to the following reasons:

1. The antibiotic resistance markers identified as markers for creation of GM plants are naturally occurring. nptII gene is found in human intestinal bacteria. The process of genome transfer is also more common and easier making it a suitable candidate for such studies.
2. The antibiotics in question, neomycin and kanamycin are seldom used for therapeutic purposes and less toxic ones have been developed.
3. Gene transfer mechanism between the GM plants and soil bacteria are not fully understood and proved in practicality. There have been no known instances of soil bacteria taking up the plant DNA in nature and making functional copies of the said genome.
4. Plant DNA irrespective of whether it is genetically modified or not ends up the gut of humans as a substrate for the pancreatic fluids and gets degraded in the acidic environment of stomach.
5. Gene transfer frequency is lower or practically nil between the plant and bacteria. Gene exchanges and horizontal transfer among the bacterial strains and species are relatively higher and account for most of the development of acquired resistance among microbes.
6. Antibiotic resistance marker genes only offer resistance against effect of antibiotics. They do not code for the production of the antibiotics and hence GM plants with antibiotic resistance gene markers do not carry antibiotics in their crops.

Nevertheless, there are major issues associated with the creation of such GM plants. Development of Antibiotic resistant superbugs is the most prominent one. First generation antibiotics are almost becoming obsolete due to the emergence of such superbugs. For example, multiple antibiotic resistance is more common among the microbes causing hospital acquired infections and diseases such as Tuberculosis.

Phasing out GM crops with antibiotic resistance markers have been therefore widely accepted in most countries of the European Union. The concerns of antibiotic resistance and emergence of superbugs have been raised since the 1990s. The Norwegian parliament in June 1997 decided to prohibit the production, import as well as marketing of Genetically Modified Organisms with antibiotic resistance marker genes and pledged to actively advocate for the cause among the international audience.

Developing countries often lack a rigid biosafety regulation framework due to which the consequences of such genetic experiments are often irreversible and damage the native biodiversity of the country. Also, there is a need for more concerted studies of effect of introduction of such Antibiotic resistant plants in the environment as well as in relation to the humans and animals in the food chain.

Marker genes are not normally removable from the transgenic organism including plant cells. Hence the recent research focus has been primarily on the methods to remove the marker genes. Several methods are under extensive study. The most prominent ones make use of

1. Meganuclease enzyme mediated removal of marker genome. It makes use of specific recognition sequences which are introduced on either side of the marker gene and serve as an identification spot for the enzyme to act upon. The system cannot be adopted in those plant cells whose genome already carries such recognition sequences since the enzyme can damage the plant DNA itself.

2. Researchers have tried to insert the marker gene on one DNA construct and the trait on another one. They integrate to different chromosomes when integrated into a plant genome. The transfer of transgene to the next generation can therefore be controlled and the frequency of transfer also varies. Controlled regulation of the gene transfer frequency has however been nearly impossible to achieve and hence need more efforts.

3. Homologous DNA sequences can alternatively be used to mark the presence of antibiotic resistant marker genes. These sequences can undergo random recombination and subsequent elimination. But the frequency is very low and is specific to the plants.

There are several studies which confirm the fact that the genetic material of the transgenic organisms ingested as food are not easily degraded and there is a possibility of gene transfer to the gut bacteria as well as the mammalian cells. However, the effects were not of uniform pattern and hence not been quantified accurately to establish a significant relation to have a blanket ban on transgenic organisms including genetically modified plants.

Recent studies on the emergence of antibiotic resistant microbes have revealed the primary cause to be the indiscriminate use of antibiotics both by humans and as constituent of animal feeds. The direct contribution towards this by the GM plants seems to be quite negligible. However, a recent effort by the Danish Government to ban Antibiotics in growth promoters used in animal feeds have backfired with a sudden increase in livestock diseases and the use of therapeutic antibiotics shot up to 223 %. It has also resulted in resistance to antibiotics other than neomycin and kanamycin. Hence, systematic studies addressing concerns on safety and toxicity are the need of the hour rather than introducing sudden and blanket ban on the use of antibiotics as well as antibiotic resistance markers.

5 EXAMPLES OF GM CROPS AND THEIR SIGNIFICANCE

Crops are genetically modified for the various reasons we have analyzed in the previous chapters. However, the major priority of research points to the primary factor of food adequacy. This is of significance since the area of cultivation has been generally on the decline especially in many tropical and subtropical countries.

Consider the global warming and increased emissions of greenhouse gases. The scenario is well addressed by a study done by Zhang and Cai in 2011. They proposed two scenarios in which one follows the general pattern of increase in Carbon dioxide levels and the other considers a worldwide adoption of greener economy. Although the variables used are much ambiguous and subject to random variations, the analysis have revealed a general grim reality that the world is sure to meet sooner.

Though the green economy initiatives by the various countries contribute to a rise in the agricultural land especially in the higher latitudes, the food production might not be able to sustain the whole population considering the current model of human population growth. Add to this the scarcity of water for cultivation, conversion of agricultural lands to human settlements, issues in natural conservation and ecological disasters; the net available arable (cultivatable) land is surely going in a declining trend.

The current practices of agriculture hence need to be modified to meet the demands of the human population through eco-friendly and ethical means using technology. The rise of GM crops in the past century has been such an initiative by the scientific community.

1. GM RICE

Rice has been the most widely cultivated food grain and the major source of carbohydrates in diet of many developing and underdeveloped countries around the world. However, with rice being the main ingredient in diet, malnutrition is a serious problem affecting these countries.

Approximately, 3 billion people around the world have rice as the main staple diet. Although the rice plant (oryza) does produce the plant pigment Betacarotene (precursor of vitamin A), the pigment is limited to the outer grain layers of rice seed. This layer is removed during the milling and polishing process used to enhance the shelf life of the rice grains facilitating their transport and storage. The inner endosperm portion lacks Betacarotene and the rice we include in the diet hence does not carry the benefits of vitamins in diet. The traditional methods of crop modification including hybridization do not work since all species of oryza lacks genome to produce Betacarotene in endosperm.

Vitamin A deficiency is a serious concern in the low income countries with a high prevalence of 1.5 per 1000 children. In simple terms, approximately 75 % of the total blind children around the world live in the low economic regions of Asia and Africa.

GM rice was the first initiative to address this concern to alleviate the effects of malnutrition in the low income countries. A strain of the oryza with high Betacarotene was developed by the scientists at Swiss Federal Institute of Technology for Plant Sciences called as 'Golden Rice". Gene for phytoene synthase (psy) was isolated from the wild daffodils ((Narcissus pseudonarcissus) along with phytoene desaturase (ctrI) gene from a bacteria Erwinia uredovorainto and inserted into the Oryza sps Genome enabling the production of Betacarotene. The fortified kernel is golden yellow in colour.

However, the effort failed to reach the public at first due to the widespread patenting of the process involved (around 70). The purpose of 'Golden Rice' as a humanitarian aid to alleviate malnutrition soon got entangled in the intellectual property rights war of patents, especially involving corporate giants such as Syngenta (formerly Zeneca) and Monsanto holding key patents. The obstacles were somewhat side-lined with the intervention of international legal expertise, scientific community and especially in the interests of the two men behind the development of golden rice, Ingo Potrykus, Professor of Plant Science at the Swiss Institute of Technology and Peter Beyer, Professor at the Department of Cell Biology at the University of Freiberg.

The Golden rice soon reached the IRRI (International Rice Research Institute) in Philippines for further research. The outdoor field trials were carried out in 2005 in U.S. However, since the adverse effects of Genetic Modification have not been precise, there have been widespread doubts and fears which catapulted to for massive public opinion against GM crops. Organizations such as Greenpeace dismissed the GM crops as disrupting the natural ecology and unethical.

The idea of opposing GM rice was vehemently supported by the claim of nutritionists that intake of Vitamin A needed lots of fats in the diet which the poor man's staple diet Rice lacked completely. In the absence of a diet with fats, the intake of vitamin A through rice was practically not feasible and hence the total experiment, a failure. Also, with the average consumption of rice of an adult Asian, only 8 % of the daily dose of requirement of vitamin A could be met even with the introduction of Golden Rice.

Golden Rice version 2 was developed in 2005 with a substitution of maize DNA in place of daffodil DNA which increased the provitamin Betacarotene amount to 23 times the original. But the concerns of Vitamin A absorption could not be solved.

Ecological concerns have also surfaced whether such GM plants could cause genetic contamination altering the natural ecosystems and the effects of wind pollination among the GM crops in the spread of transgenes.

The European Union has been swift to adopt a precautionary approach in the implementation of cultivars without establishing the determined effects of GMOS in food production. These technical and ethical concerns have in turn forced the IRRI to shift the focus towards enrichment of rice species with other minerals such as iron, and development of plants with enhanced photosynthetic ability to increase the yield.

2. GM Potato

The Genetically modified potato known as New Leaf potato was developed by using the genes from the soil bacterium Bacillus thuringiensis to provide resistance from the potato beetle (Leptinotarsa decemlineata) infection. Monsanto and co took up the marketing of the potato variety in 1990s, but was soon withdrawn in 2001 after widespread protests against GM crops.

Another variety Fortuna Potato developed by BASF Corporation by inserting two genes from wild potato strain *Solanum bulbocastanum* resistant to late blight infection. Late blight infection is caused by *Phytophthora infestans,* affecting over 20 % of the annual yield of potato. Remember the Irish potato famine of 1840s! Phytophthora infection was the primary cause for the famine. After much legal hassles, Fortuna failed to obtain approval for field cultivation in Europe which forced BASF to abandon its efforts.

A second variety resistant to late blight infection was developed by British researchers known as "Desiree Potatoes". The field trials have shown promising result of increasing the yield to almost 100%. The Desiree potato hopes to find a market in U.S and is associated with Simplot for further research on improving its immunity.

Since the GM food found it hard to find to get approval for direct human consumption, much of the research efforts have been to meet industrial product demands. For example, "Amflora" potato was developed by BASF for production of pure amylopectin which can yield waxy potato starch. Amflora was developed mainly for industrial applications like paper making. Although European Union approved the use of Amflora for industrial purposes in 2010, the permission was withdrawn after two years due to widespread opposition. BASF continues to do research on improving the varieties hoping to find suitable markets in America and Asia.

3. GM Tomato

Ripening of tomatoes under natural conditions is not practically feasible since they get softened during the process and make them unsuitable for shipping. Normally, they are picked by hand while unripe and the usual ripening process while shipping does not allow them to develop the natural flavors. Artificial methods of ripening using ethylene gas have been resorted to in most cases. Although this method lacks potential health hazards since ethylene is a natural hormone found in plants responsible for ripening, the natural flavors and aroma was lacking. And they could not be stored along with ethylene sensitive vegetables like cabbage, broccoli etc.

This led the researchers of Calgene to develop Flavr Savr using antisense technology. The enzyme polygluconase production was slowed so that the fruit will ripen slowly maintaining the stiff skin. Unlike the traditional tomatoes, these could be allowed to ripen on the vine for longer duration due to firmer skin and therefore can be transported with ease after ripening.

The GM tomatoes were found to have higher concentration of pectin molecules which accounted for higher serum viscosity. But they were in no way inferior to normal tomatoes in terms of pH, acidity, color and sugar concentration. The only notable difference was the lack of polygalacturonase activity which was responsible for hydrolysis of pectin molecules. This difference contributes to the notable changes in quality of fruit such as enhanced shelf life, firmer fruits even after ripening, and improved resistance to fungal pathogens in Flavr Savr. But the public opposition was strong and hence the 'Flavr savr' soon found its place off the shelves.

Transgenic tomatoes have since then developed with therapeutic effects. For example, those developed by Lopez et al with a transgene for interleukin-12 has been found to be therapeutic beneficial to treat pulmonary tuberculosis in animal studies. Similarly, studies done by Konijeti et al have revealed the tomatoes to exhibit chemopreventive effects when a small structural change was introduced in the natural lycopene. Genetic modification to alter lycopene structure might therefore be a possible area of further research in chemoprevention. Similarly, Tomato leaf curl virus resistant plants were developed using Agrobacterium mediated Genetic modification by S.K. Raj et al in 2005.

4. GM Papaya

The 1990s witnessed a major disaster which affected the Hawaiian Papaya sales due to the deadly ring spot virus. Rainbow Papaya developed in the 1998 proved to be resistant to the virus and since then almost 80% of the Hawaiian Papayas under cultivation are genetically engineered. The ring spot virus would have otherwise wiped out the crop itself since no known natural or organic farming methods could prevent the infection.

5. GM Strawberry

Strawberries generally are rich in vitamin C content. Researchers have successfully transferred the yA gene (GalUR) which converts the plant protein to vitamin C from cress Arabidopsis thaliana to Strawberry to enrich the vitamin content to almost 3 times. Agrobacterium mediated transfer of the gene to the plant resulted in over expression of the gene resulting in higher vitamin C content.

Strawberries, like tomatoes encounter problems of ripening while transport and similar researches have been undertaken to prolong the shelf life. They are found mainly in the temperate lands with cold temperatures and are frequently affected by frost. Frost damage causes a lot of loss in the final yield and is an ultimate price control factor in the market. Transport temperature conditions also should be taken care since the frost conditions can cause damage.

Preventive measures against frost which uses large scale coverage of farms and wind machines are less effective. Scientists have therefore tried to reduce the damage by development of transgenic strawberries which are resistant to frost.

Antifreeze proteins found in fish (Antarctic nototheniid fish) protect the fish from damage due to freezing. The genes for AFPs are found in many other plants, fungi etc. also. Although the genes for AFPs were isolated and transformed cell lines were obtained using Agrobacterium mediated transformation, the rate was very low for new saplings to be successfully grown into full-fledged plants. The research is still in its infancy and might take longer time for a successful frost resistant strawberry to be produced by GM technology.

6. GM Corn

Corn is one of the few crops that have made its way to supermarket shelves along with GMO label. Most of the corn harvested in U.S has undergone genetic modification. But much is used for animal feed or for production of ethanol. GM sweet corn has been approved for cultivation in U.S since mid-1990s and the widespread cultivation has resulted in almost all the corn harvested to be practically genetically modified. This is aggravated by the cross pollination and wind dispersal which is common in maize (corn plant).

Genetic modification in Corn has been mainly to induce herbicide and pesticide tolerance and as a protective measure against disease causing insects. Bt Corn is one such example which is intended to control the population of lepidoptera larvae which causes corn borer infection. However, more than the inserted Bt gene, the promoter and identifier sequences of genetic material which are added for selection of transgenic plants have proven to give rise to more harms often unexpected. How these promoter sequences code for in a new genome is not well understood and now the primary contention of those who oppose GMO foods.

GM corn developed by Syngenta have usually one or more of the genes like Cry1AB(resistance to lepidopteran larvae), PAT (phosphinothricin-N-acetyltransferase inducing herbicide tolerance),EPSP synthase (glyphosphate resistance), mCry3A (against root work infestation) etc. Specific corn varieties coding for enhanced alpha amylase enzyme enabling its use as substrate for ethanol and high fructose corn syrup production also has been developed.

7. GM Canola

Major portion of the Canola oil produced in U.S is now from genetically modified canola plants whose gene structure has been altered by using transgenic technology to induce herbicide resistance or pest resistance. Genetic modification of canola has also been carried out to improve the oil composition so as to improve the non-triglyceride portions which are usually removed from the oil during refining process. These remains are used to produce animal feed.

8. GM Zucchini, GM Sugar Beet etc.

Virus resistant GM Zucchini constitute approximately 13 % of the total zucchini produced in U.S. Herbicide tolerant species of sugar beet is approved and is cultivated in many countries including U.S, Mexico, Australia, New Zealand, Canada, EU, Japan, Philippines, Korea, Singapore and Russia.

9. Bt Cotton

The lepidopteran larvae resistant GM cotton variety was

introduced in U.S in 1995 for commercial use. The Bt cotton was soon introduced to other countries including India by Monsanto and Mahyco. By 2014, India has emerged as the single largest producer of GM Bt Cotton and over 95 % of the total cotton cultivated in India is genetically modified. The Bt cotton with the introduction of Cry gene is able to prevent infestation by three types of bollworms. The growth of Bt cotton has also significantly reduced the use of insecticides.

Integrated methods of pest management could also be effectively implemented due to the lowered use of pesticides and insecticides. However, Bt cotton has had its share of limitations including the high cost of seeds, the short duration of its efficiency to produce the Bt endotoxin (up to 120 days), and the ineffectiveness against pests other than the bollworms.

The higher cost has led to other players to try developing cheaper varieties and those with more quality seeds. Punjab Agricultural University has succeeded in developing cheaper alternatives to Bt cotton prompting the Indian government to lift a short ban which was imposed on GM cotton.

The Bt cotton however continues to dwell in controversies aggravated by the high seed cost, multinational company monopolising the seed markets and resultant farmer suicides. The benefit of high yield is yet to become profitable for the farmers.

Some Myths Associated With GM Crops

Although there has been much public outcry against the production and commercialization of GM crops, especially those for human and animal consumption, GM crops tend to get to our dishes most of the time disguised under an incomplete labelling.

But there are many other crops which are thought to be genetically modified, but are not actually.

For example, GM Potatoes and Golden Rice have been withdrawn from the market after consumer protests. The seedless watermelon, which many consider as genetically modified is but a hybrid strain only. Meat, fish and eggs have not been genetically engineered so far although some researches are being conducted to develop the same.

Intentional efforts by various governmental organizations to keep the GMO foods under organic foods have been also withdrawn due to public concern and new specific standards for organic food have been drafted by the USFDA.

And there is the myth that all genetically modified plants are crops. Not essentially, some like Poplar trees are genetically modified to enhance sequestration of heavy metals from ground water as a phytoremediation strategy. Such efforts could somehow help in the clean-up of contaminated soil and water resources to a certain extent and hence should be appreciated.

REFERENCES

1. Guidance for Industry Regulation of Genetically Engineered Anim als Containing Heritable rDNA Constructs; U.S. Food and Drug Administration; http://www.fda.gov/cvm/Guidance/guide187.htm

2. Primose, S.B. and Twyman; Principles of Gene Manipulation and Genomics; 7th ed. 2006. Blackwell Publishing, Malden, MA.

3. Lundmark, C.; Genetically Modified Maize. Bioscience 57 (11): 996; December 20 07

4. Widhalm, S. January 2006. Pros and Cons of Tinkering With Crop Genetics. World & I 21 (1):8

5. DeynzeA. 2004. Roundup Ready Alfalfa: An Emerging Technology . Available at: anrcatalog.ucdavis.edu/pdf/8153.pdf

6. BabiliS, KlotiA, Zhang G, Lucca P, Beyer P, PotrykusI. January 14, 2000. Engineering the ProvitaminA(BetaCarotene) Biosynthetic Pa thway into (CarotenoidFree) Rice Endosperm. Science 287 (5451): 303

7. HancookR. March 15, 2006. Improving the Nutritional Value of Cr ops by Genetic Modification: Problems and Opportunities Illustrate by Vitamin C. Asia Pacific B iotech News 10(5): 237

8. WHO Food Safety Unit. Health aspects of marker genes in genetically modified plants, 1993. Report of a WHO Workshop.

9. C. Zhang et al. Genetically modified Foods- A critical review of their promise and problems; Food Science and Human Wellness 5 (2016) 116–123

10. American Medical Association. "Genetically Modified Crops and Foods." December 2000. Available at <http:// http://www.ama-assn.org/ama/pub/article/2036-4030.html>

11. Antibiotic Resistance Markers in Genetically Modified (GM) Crops; Briefing paper no 10; EUROPEAN FEDERATION of BIOTECHNOLOGY TASK GROUP ON PUBLIC PERCEPTIONS OF BIOTECHNOLOGY

12. Klümper W, Qaim M (2014) A Meta-Analysis of the Impacts of Genetically Modified Crops. PLoS ONE 9(11): e111629. https://doi.org/10.1371/journal.pone.0111629

13. Prabhuprasad Paduchuri, Transgenic tomatoes – A Review; International Journal of Advanced Biotechnology and Research ISSN 0976-2612, Vol 1, Issue 2, Dec-2010, pp 69-72. http://www.bipublication.com

14. Parliamentary Postnote 360 June 2010 Genetically Modified Insects; The Parliamentary Office of Science and Technology, 7 Millbank, London.

15. Mohamed N. Sallam ; INSECT DAMAGE: Damage on Post-harvest; Post-harvest Compendium; FAO- United Nations

16. Deborah Whitman: Genetically Modified Foods; 2000; CSA DiscoveryGuides http://www.csa.com/discoveryguides/discoveryguides-main.php

ABOUT THE AUTHOR

Sandhya holds a Post Graduate Degree in Biotechnology and is a writer by profession. She writes on various topics primarily concerned with Science and Nature. Her major interests are reading and statistics. She likes to raise awareness about ecological conservation and biodiversity.

Sandhya Anand